GRIN Publishing

Geology Adventures

James Tallant

Bibliographic information published by the German National Library:

The German National Library lists this publication in the National Bibliography;
detailed bibliographic data are available on the Internet at http://dnb.dnb.de .

Imprint:

Copyright © 2007 GRIN Verlag, Open Publishing GmbH
Print and binding: Books on Demand GmbH, Norderstedt Germany
ISBN: 978-3-640-84021-2

This book at GRIN:

http://www.grin.com/en/e-book/167473/geology-adventures

GRIN - Your knowledge has value

Since its foundation in 1998, GRIN has specialized in publishing academic texts by students, college teachers and other academics as e-book and printed book. The website www.grin.com is an ideal platform for presenting term papers, final papers, scientific essays, dissertations and specialist books.

Visit us on the internet:

http://www.grin.com/

http://www.facebook.com/grincom

http://www.twitter.com/grin_com

Geology Adventures

James Tallant

Axia College

SCI 245 Physical Geology

August 19, 2007

Welcome to Geology Adventures 2007. Our tour will have five stops showing three different types of geological features. These will include a volcano or volcanic activity, a coastal feature and a ground water feature. At each stop, a brief description will be given on the features regarding how it was formed, the characteristics it has, interesting facts, myths, natural or recreational opportunities and any special precautions that are necessary. The itinerary for this tour will begin in Kilauea, Hawaii; then moves onto Panama City Beach, Florida; and finally ending our tour at Yellowstone National Park's Old Faithful geyser.

First, before we begin our tour of these sites a brief discussion of the geological events and the importance of geology in our daily lives must be examined. Geology itself is the study of the earth's natural aspects by using scientific methods (Plummer, 2004). In understanding geology, scientists and geologists can explain how the earth's surface and interior are constantly changing. The resulting changes create the world's natural resources, beauty, and at times its natural disasters.

As geologists uncover the scientific answers to some of the mysteries of the earth's secrets, they can predict future events that may be disastrous to populated areas or predict locations of valuable natural resources. According to Plummer (2004), Geology combines the interaction of "the atmosphere, water and rock, the modern theory of plate tectonics, and geologic time" to form the overall larger big picture of how the earth's workings occur. Geology, the study of the earth's natural resources benefits everyone. The natural resources of the earth supply things we need from fuel, water, rocks, sand, and gems. Natural resources are not limitless. Without understanding how these resources came about and how to conserve them, the future quality of civilization is questionable.

Whether searching for new oil deposits or predicting the next volcanic eruption geologists are constantly studying the earth looking for signs of abnormity. Geology and the resulting natural resources of the earth supply things that we as humans need. Because of discovering and extracting these resources, geologists have been given the task of protecting the environment at the same time. However, understanding the geological forces that created the natural resources will assist in minimizing the geological hazards that can result from extracting them.

The earth's continents and oceans sit atop moving plates. These plates are in constant motion. Movements of these plates are the beginning of some of the earth's most powerful natural occurrences including earthquakes, volcanoes, and tidal waves. As these plates move, escaping magma moves toward the surface of the opening. As a result steam, vapor, and molten rock attempts to escape. Over a period of time, this repeated process creates additional land mass.

We will begin our first stop of the tour visiting volcanoes and volcanic activity. "Kilauea the world's most active volcano, on Hawaii's big island has been erupting continuously since 1983" (Weekly Reader News, 2006). Hawaii, part of the United States, is a popular vacation resort retreat located in the mid-Pacific Ocean as part of an island archipelago chain. These islands were formed from underwater volcanoes. The Hawaiian Islands are a series of layers of basalt from shield volcanoes built upwards from the ocean floor by intermittent eruptions over millions of years.

Kilauea is constantly active. Currently, according to the United States Geological Survey (USGS), Kilauea itself has not had any new activity since the June 18-19, 2007 eruption. Recent tiltmeter measurements have indicated that Kilauea's north side, Pu' u' O' o, tilt has increased indicating that the floor of the crater is continuing to collapse. There has been a decrease in seismic activity and very few earthquakes recorded. These indicators point to a return to pre June 17, 2007 levels.

Kilauea, along with all Hawaii's volcanoes is spectacular to observe. These eruptions are not usually violent or dangerous because their lava flows are less viscous or fluid. Hawaii has two types of surface basalt lava flows. The first type is pahoehoe which is ropy or billowy surface formed by the quick cooling and solidification from the surface downward of a lava flow or a lava pool that was fully liquid (McGraw-Hill, 2004). The second type of flow is basalt, which has cooled enough to have partially solidified and moved forward as a pile of rubble and is classified as a flow. A minor feature of lava is created from spatter cones. These are steep-sided cones built from lava sputtering from vents, which will occasionally develop into a solidifying lava flow.

The constant eruptions have a beneficial side. These non-violent volcanic eruptions have been beneficial to the state of Hawaii's tourism economy. These erupting volcanoes are relatively safe and the exceptional spectacles attract tourists and scientists alike which is a benefit to the islands overall economy. Although these eruptions are usually non-violent, homes are occasionally destroyed. The result of constant eruptions and lava flows to the sea and then solidifying has added land mass to the Hawaiian Islands. The Hawaiian Islands are literally growing in physical size. Another benefit of the constant eruptions is the weathered volcanic ash and lava. These produce fertile soil in which pineapples and papayas are grown.

Myths and religions flourish in cultures that live with volcanoes. Hawaii's, Madame Pele, is regarded as the goddess who controls eruptions. According to legend, Pele, and her sister tore up the ocean floor to create the Hawaiian Islands. These eruptions usually begin with lava gushing from a long crack in the earth leaving a smooth skin like texture from the fluid trail. Hawaiians regard this as symbolism of a woman's vagina, and her menstrual cycle (Dvorak, 2007). Pele, with the red lava always heading to the sea is the same path that ancient Hawaiian women took to cleanse themselves. Even today, many people strongly believe that Pele dictates when and how severe an eruption will occur.

Kilauea's constant eruptions have little recreational value other than just to behold the awe-inspiring power and beauty of an eruption. The eruptions give scientists more valuable data and information regarding the inner workings of earth's inner structure. With this information, scientists can provide better warnings for future eruptions to minimize loss of life and property as well as explore future fuel resources such as geothermal energy.

Kilauea is a shield volcano that has been constantly active since 1983. The eruptions have been beneficial for tourists and scientists alike. The islands of Hawaii will continue to benefit, and literally grow because of Kilauea, and her sister volcanoes eruptions in Hawaii. Each successive eruption provides valuable data regarding the questions of how volcanoes begin and the age of the earth and rock cycles. The eruptions also instruct us in the importance of respecting nature and keeping a safe distance from them. Geology will continue to explore and research the inner workings of volcanoes for years to come.

Our next stop on Geology Adventures 2007 brings us to the home of "the world's most beautiful beaches" (The city of Panama City, 2007). Panama City Beach is an example of a beach and coastal feature. As with the previous stop, a brief description will be given on the features regarding how it was formed, the characteristics it has, interesting facts, myths, natural or recreational opportunities, and special precautions that are necessary.

Located in the Northwestern part of the state of Florida, Panama City Beach overlooks the Gulf of Mexico. Panama City's average elevation is 13 feet above sea level. Panama City Beach is part of a chain of beaches along the Western Gulf Coast known as the Emerald Coast or the Redneck Riviera (Wikipedia, 2007). This beach is a depositional coast, which is a gently sloping plain that shows minimal effects of wave erosion. According to Finkl, Andrews, and Benedet, "Sediments off the southwest Florida coast are part of a larger continuum that lies at the center of an ancient carbonate platform that faces an enormous ramp" (2006), which influences coastal availability of sediment and wave energy.

Over many years, the longshore drift of sand around the peninsula of Florida and into the large bay or Gulf of Mexico created the beautiful beaches along the coastline including Panama City Beach. The coastline is constantly in a state of regression and expansion. Natural events of wave refraction, hurricanes, and climate changes such as El Nino change the face of the beach. Human interference also contributes to beach changes by building more piers, marinas, and beachfront homes. Each of these events changes the landscape of how waves reach the shoreline and distribute sediment. If left in their natural state, a shoreline beach can self sustain itself through cycles of weather and environmental changes. However, human intervention creates obstacles that nature cannot overcome. Waves find a way around human built obstacles and thus deposit sediment in greater amounts and in fewer amounts than they otherwise would.

Panama City Beach is a year round tourist attraction. Panama City has the beauty of its beaches as well as deep-sea fishing, offshore boat racing, clubs, amusement parks, and state parks. St. Andrews State Recreation Area is one of the most popular outdoor recreation spots in Florida (The city of Panama City, 2007). The beach is best known for its one and a half miles of white sand beaches, natural dunes, and crystal clear water (The city of Panama City, 2007). St. Andrews bay and Panama City legend has it that many pirates used the deep pockets of the bay to await the heavily laden Spanish gold ships leaving Mexico and buried their treasure at Spanish

Shanty Point (The city of Panama City, 2007). Another interesting fact regarding Panama City

Beach, is that during World War II, it was part of the St. Andrews Sound Military Reservation.

Circular canon platforms are still in place near the jetties.

The future of this beach depends heavily the development of the surrounding areas. Increased

human population and the resulting building of beachfront properties, marinas, and piers will

artificially change the wave the beach naturally takes care of itself. In most cases human

interference, even if under good intentions, usually ends up doing more harm than good. Off

shore oil drilling is a threat as well and is damaging the ecosystem environmentally and

aesthetically. The beaches are further burdened as it receives more human made pollution that

makes its way toward shore from the ocean. Pollution generated from other areas inland or ocean

bound are delivered by rivers and dumped in to the Gulf of Mexico will occasionally surface

along the shoreline of Panama City Beach (Unfar, Unfar, Ellender, Rebarchik, Stone, 2006).

A balanced approach to maintain the beauty of Panama City Beach with the encroaching

development of humans will be necessary. Development must be kept under control by

minimizing artificial building along the coastline with beachfront homes, marinas, and piers. In

addition, the effect of storms such as hurricanes, which are at first devastating on the surface are

eventually countered by other natural occurrences that keep nature in balance. All this with

minimal human intervention will keep Panama City Beach as beautiful tomorrow as it is today.

The last site to be visited on Geology Tours is the ground water feature Old Faithful at

Yellowstone National Park. Old Faithful is a cone geyser located in Yellowstone National Park

in Wyoming (Wikipedia, 2007). Yellowstone itself sits on a large caldera or a large circular

depression at the summit of a volcano. Old Faithful was the first geyser named at the park by the

Wasgburn-Langford-Doane Expedition in 1870. Old Faithful is not the tallest or largest geyser at Yellowstone. However, Old Faithful is one of the more predictable geysers.

Old Faithful and Yellowstone National Park rest on a unique geological spot called a hydrothermal area. The park is located at the intersection of the Basin and Range Province and the Snake River Plain. The park contains over 10,000 geysers, hot springs, and mudpots. Other geological interests include recent volcanism, glaciation, and many water falls (Yellowstone National Park, 2007).

A geyser is a hot spring that tosses underground water into the air (Yellowstone Media, 2007). A geyser begins its development by falling water or snow that slowly seeps through porous layers of rock. Eventually the water encounters extremely hot rocks that have been heated by magma, a large body of molten material under the park. The hot water rises through cracks and fissures as a natural plumbing system of a thermal water feature. These water shots can heights of 184 feet and last up to 5 minutes (Yellowstone Net, 2007).

Old Faithful is one of the more predictable geysers in the world. However, there are some common misconceptions. Some of these include that the geyser plays or erupts every hour and on the hour. This is a misconception dating back to the original Washburn expedition where one of the members had reported that the geyser played nearly hourly. Other misconceptions include the thought that the geyser is slowing down in speed and shooting lower water heights. These to are ill-founded and subject to human error. Overall, according to scientific data Old Faithful's average interval has remained within a few minutes of its previous times.

These three distinctive geological features have similarities and differences. The similarities arise between the volcano feature and the ground water feature. Each is controlled by the earth's deep workings. As hot molten magma, is either forced upward or placed in contact with another material under ground by way of the movement of continental plates a volcanic or geyser eruption will occur. In rare occurrences a coastal feature can be affected by plate tectonic movements under the ocean by way of an under sea volcanic eruption that creates a tsunami that makes landfall by way of a coastline. The coastline will be adversely affected by erosion of sediment in land and out to sea. The water that surrounds them mainly affects coastal features. The water or ocean in many cases depends on the wind to generate its energy. As water waves are generated out at sea, they move toward land and depending on the coastline and depth of the sea floor along the coast dictates how and where sediment will be moved creating unique coastal features.

On the other hand volcanic activity relies on the forcing of magma and lava flows out of its openings and changing the landscape as it moves. As the pressure and temperature of hot molten magma reach a certain point then it forces it way to the surface. As it cools, it turns into rock. Geysers rely on heated magma to boil underground water then force its way upward though cracks in the rock to form narrow shots of water into the air. These small ground water features in themselves are not necessarily dangerous. They do represent what is below ground, which is a large volcanic area that taken in the entire context of Yellowstone National Park can be dangerous.

Geology will continue to emerge as an important part of science. The study of the earth's internal workings and understanding how resources can be found and conserved will be the forefront of geologist responsibilities for years to come. Geological events happen for a reason. Whether a natural disaster, a new oil deposit is found or a spectacular view in nature, there is an underlying geological answer. Geology must remain on the forefront of everyone's minds as it holds the key for the future of energy development and conservation, which ultimately is the key

for future human survivability.

References

Dvorak, J. (2007, January). Volcano Myths and Rituals. *American Scientist*, 95(1), 8-9. Retrieved June 24, 2007, from Acedemic Search Premier Database.

Emerald Coast. (2007). In *Wikipedia* [Web]. Retrieved July 22, 2007, from http://en.wikipedia.org/wiki/Emerald_Coast

Finkl, C. W., Andrews, J. L., & Benedet, L. Assessment of Offshore Sand Resources for Beach Nourishment Along the Southwest Coast of Florida. *Coastal geology and Geomatics*, Retrieved July 22, 2007, from http://www.fsbpa.com/06Proceedings/04-Charles%20W.%20Finkl%20%20Jeffrey%20L.%20Andrews%20and%20Lindino%20Benedet.pdf.

Geology. (2007). Retrieved August 12, 2007, from Yellowstone National Park Web site: http://www.yellowstonenationalpark.com/geology.htm

Kilauea's Fiery Show. (200, January 27). *Weekly Reader News – Edition 4*, Retrieved June 24, 2007, from MasterFILE Premier Database.

Old Faithful Geyser. (2007). In *Wikipedia* [Web]. Retrieved August 12, 2007, from http;//en.wikipedia.org/wiki/Old_Faithful_Geyser

Plummer, C. C., McGeary, D. and Carlson, D. H. (2004). *Physical Geology* (10th ed.). New York: McGraw-Hill.

The city of panama city beach, (2007). Visitors Area Information. Retrieved July 22, 2007, from The Official Site of the City of Panama City Beach, Florida Web site: http;//www.pcbgov.com/visitors.htm

The geysers of yellowstone. (2007). Retrieved August 12, 2007, from Yellowstone National Park by Yellowstone Net Web site: http://www.yellowstone.net/geysers/oldfaithful.htm

Yellowstone national park. (2007). Retrieved August 12, 2007, from Geology Field Trip Across

the Western USA Web site: http://geology.csub.edu/vtrips/ynp/Yellow.htm

Ufnar, D., Ufnar, J., Ellender, R., Rebarchik, D., & Stone, G. (2006, November). Influence of

Coastal Processes on High Fecal Coliform Counts in the Mississippi Sound. *Journal of*

Coastal Research, *22*(6), 1515-1526. Retrieved July 22, 2007, from Academic Search

Premier database.